D. L. C. Kumari Fonseka
Upuli Irosha Wickramaarachchi

Plant Tissue Culture

D. L. C. Kumari Fonseka
Upuli Irosha Wickramaarachchi

Plant Tissue Culture

LAP LAMBERT Academic Publishing

Imprint
Any brand names and product names mentioned in this book are subject to trademark, brand or patent protection and are trademarks or registered trademarks of their respective holders. The use of brand names, product names, common names, trade names, product descriptions etc. even without a particular marking in this work is in no way to be construed to mean that such names may be regarded as unrestricted in respect of trademark and brand protection legislation and could thus be used by anyone.

Cover image: www.ingimage.com

Publisher:
LAP LAMBERT Academic Publishing
is a trademark of
Dodo Books Indian Ocean Ltd. and OmniScriptum S.R.L publishing group

120 High Road, East Finchley, London, N2 9ED, United Kingdom
Str. Armeneasca 28/1, office 1, Chisinau MD-2012, Republic of Moldova, Europe
Managing Directors: Ieva Konstantinova, Victoria Ursu
info@omniscriptum.com

Printed at: see last page
ISBN: 978-620-2-00347-6

Copyright © D. L. C. Kumari Fonseka, Upuli Irosha Wickramaarachchi
Copyright © 2017 Dodo Books Indian Ocean Ltd. and OmniScriptum S.R.L publishing group

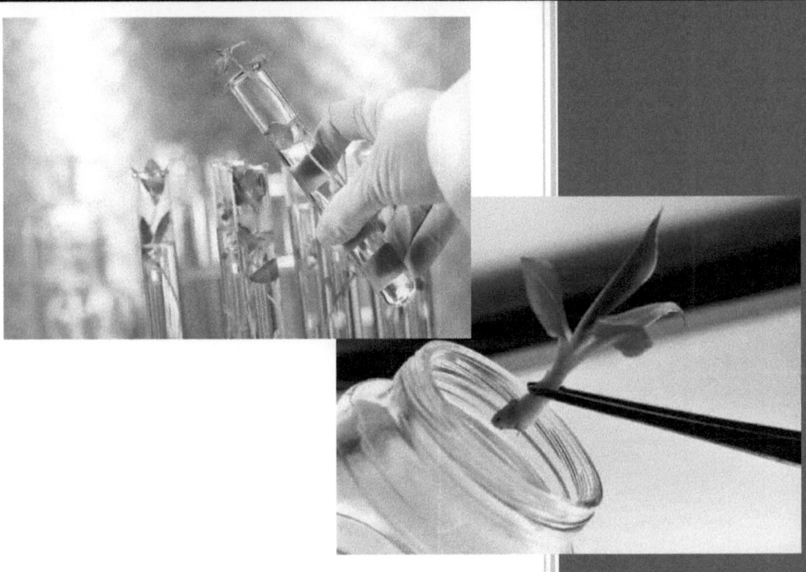

DLC Kumari Fonseka

Upuli Wickramaarachchi

Preface

A successful cultivation mainly depends on healthy and vigours planting materials, which can be considered as the most vulnerable step to initiate a fruitful plantation. Plant tissue culture technique is widely used around the world to produce high quality plants which are not able to produce via other vegetative propagation methods.

With the advances of technology and many scientific interventions plant tissue culture has developed into a vast area from production of plants similar to mother plants and the production of wide array of metabolites which have promising future in industries and pharmaceutical sector.

Initially, this was only limited to laboratory conditions, but at present farmers also use this technique to produce their own plants. Instead of costly laboratory equipment and facilities, cost effective alternates can be substituted at farmer level or small scale productions without compromising the scientific basic of plant tissue culture techniques.

This book mainly focuses on secondary, tertiary and university students, farmers and entrepreneurs who wish to have a good knowledge on plant tissue culture, a promising technology for future sustainable crop production.

Dr. DLC Kumari Fonseka
Senior Lecturer,
Department of Crop Science,
Faculty of Agriculture,
University of Ruhuna,
Matara,
Sri Lanka.
03-08-2017

Table of Contents

1. Introduction 1

  Tissue Culture Techniques...3

    Seed Culture 3

    Embryo Culture 6

    Anther Culture 7

    Callus culture 10

2. History of Plant Tissue Culture 12

3. Organization of a Tissue Culture Laboratory 14

  Basic Equipment for Tissue Culture Laboratory................................14

  Basic Sections of a Tissue Culture Laboratory 18

    Media preparation and sterilization area 19

    Transfer Area 19

    Culture Room..20

4. Preparation of Nutrient Media...23

  Inorganic nutrients..23

  Carbon energy source 23

  Organic substances 23

  Plant hormones...24

  Solidifying agents 25

  Other supplements 25

  MS Medium and Composition...26

5. Micropropagation 30

  Methods of Micropropagation 33

  Factors Affecting Micropropagation...43

6. Initiation and Aseptic Culture Establishment.................................45

7. Merits and Demerits in Micropropagation 47

Merits of Micropropagation 47

Demerits of micro propagation 48

8. Applications of Plant Tissue Culture 49

Agriculture 49

Horticulture and Forestry 50

Industries 50

Possible areas for research 50

9. Use of Low Cost Tissue Culture Materials for the Initiation and 52

Multiplication of Plants 52

What Low Cost Tissue Culture entails 52

Points where low cost tissue culture options can be applied 53

Planning a low cost tissue culture laboratory 56

10. Addressing Problems of Plant Tissue Culture 60

References 61

1. Introduction

Plant tissue culture means the *in vitro* culturing of a live plant parts in a sterilized medium within an aseptic (with no bacteria, fungi or other microorganisms) environment under controlled light and temperature conditions. It is mainly focused on the totipotency of cells which means, the ability of a plant cell to grow and develop into a complete plant when all the growth conditions are supplied at optimum levels.

Also, tissue culture technique can be used successfully to propagate plants which cannot be propagated via seeds but only by vegetative propagation techniques. Flowering plants like Anthurium, Orchid, various ornamental plants and fruit crops like banana, pineapple can be propagated easily by tissue culture technique. Using this technique, a large number of plantlets can be produced during a short period of time within a limited space.

Figure 0.1 Tissue cultured banana plants

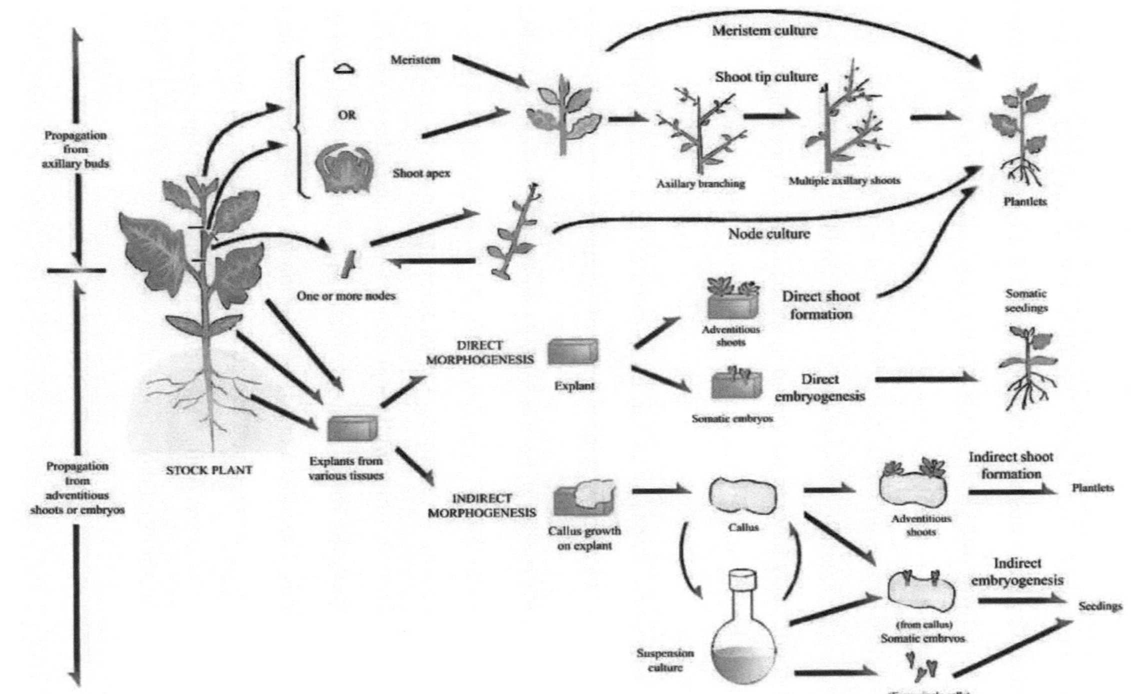

Figure 1.2 Use of different ex-plants and the mode of morphogenesis of those plant parts

2

Tissue Culture Techniques

Seed Culture

This means culturing of viable seeds in an artificial medium under sterilized conditions. Seeds produced by some plants cannot be stored for a long time due to lowering of the viability of seeds with time and ultimately the seeds lose their viability. In seeds the endosperm provides nutrients for the growing embryo. But, with time endosperm storage depletes, and it can no longer fulfill the nutrient requirements of the growing embryo, causing complete loss of viability in seeds. Under such conditions, to protect the viability of seeds, they have to be stored in an artificial nutrient medium. Then these seeds can obtain their nutrient requirements to develop into plantlets easily. For seed culture, a basic nutrient medium without hormones can be used. When seedlings appear from the germinated seeds, these can be introduced to a special medium containing Auxin and Cytokinin to obtain a large number of plantlets.

As an example orchid seeds can be considered, generally an orchid pod contains a large number of small seeds (Figure 1.3). But, most of the seeds die inside the pod before germination. Additionally, in their natural environment many terrestrial orchids maintain mutually beneficial relationships with fungi, which are required for the orchids' seeds to germinate. Many times the seeds do not contain any nutrients for the embryos, unlike almost all other seeds. The fungus provides nutrition for the embryos; in the absence of the fungus, as in the case when you are cultivating orchids, special measures must be taken to germinate the seeds. When the pod is at immature stage, these seeds should be introduced to a nutrient medium for germination.

Steps of Orchid Tissue Culture:

1. Soak the immature (green) seed capsule in 100% bleach solution for 30 minutes.

2. Dip the capsule in isopropyl alcohol or ethanol for 5-10 seconds. Remove the capsule from the alcohol   and carefully flame off the excess alcohol.

3. Under aseptic conditions, using a sterile knife or scalpel, open the capsule and scrape out the seed.

4. Carefully layer the seed over the surface of the culture medium. Seal all culture vessels.

5. It may take from 1 month up to 9 months for the seeds to begin germination. Approximately 30 to 60 days after germination begins, it will be necessary to transfer the seedlings to fresh medium for continued growth.

6. Under aseptic conditions, transfer the seedlings from the initial flasks to the flasks containing fresh medium. Place the seedlings about ¼" apart on the medium. This step is first sub culturing.

7. Allow the seedlings to continue to grow and develop. Root formation generally begins when the plants have 2-3 leaves. Continue to transfer the seedlings to fresh media every 30-60 days, increasing the spacing between the plants with each transfer/ subculture. When the flask is ready for transferring to a community pots (com pot), acclimatization process begins.

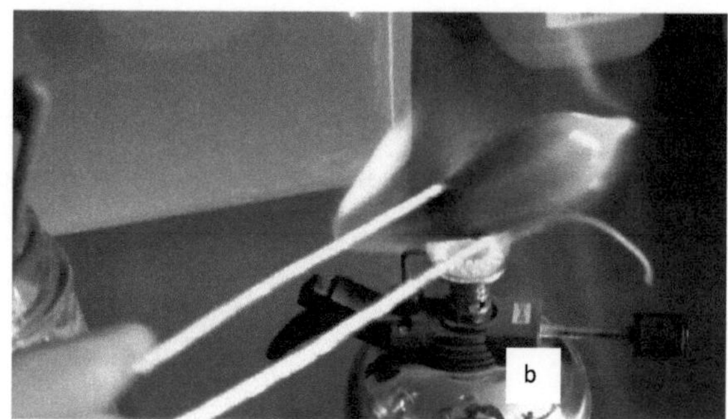

Figure 1.3 In vitro Orchid seeds germination; Orchid pod (a) Flame Sterilization of Orchid pod prior to the dissection (b) Tiny seeds inside the Orchid pod (c) and developing protocorm like bodies (d)

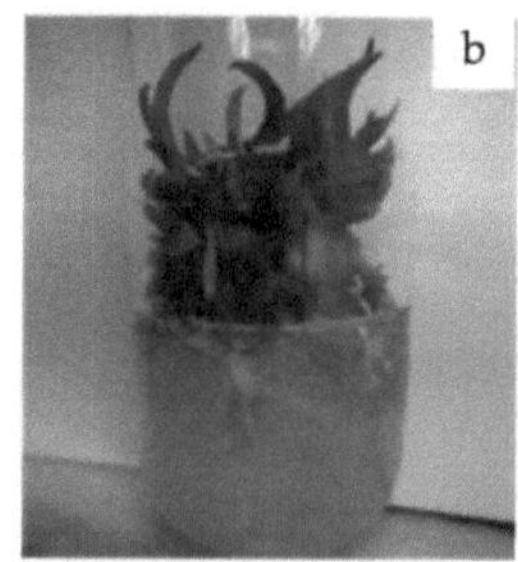

Figure 1.4 Steps (a ,b, c) of differentiation of orchid protocorm like bodies into seedlings

Figure 1.5 Orchid seedlings ready to acclimatize (a) and acclimatized seedlings(b)

Embryo Culture

Embryo from a viable seed is removed under aseptic conditions and cultured in a nutrient medium. Viable seeds produced by some plants do not germinate in field even optimum conditions are provided. This incident is known as seed dormancy. Many factors cause seed dormancy like, presence of a hard or impermeable seed coat, presence of inhibitory substances and seeds being physiologically immature. To address this issue embryo can be removed from the seed under sterile conditions and introduced in to nutrient medium or simply known as embryo rescue. Then these embryos develop into seedlings (Figure 1.6).

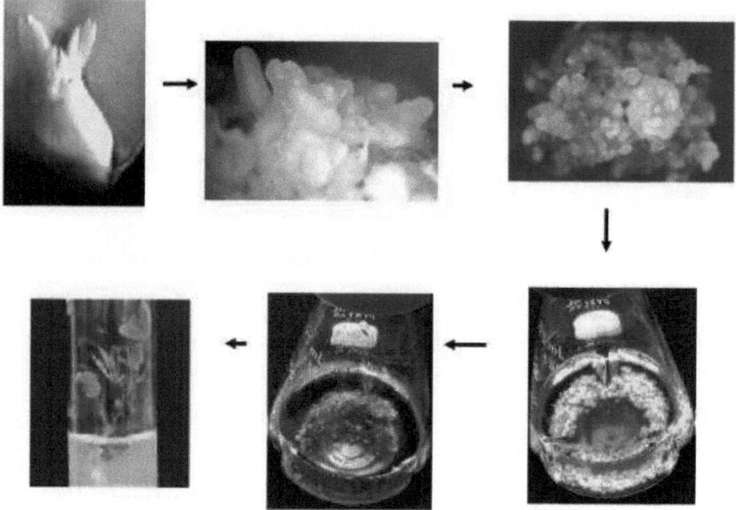

Figure 1.6 Steps of peanut embryo culturing and embryo development into seedling

Anther Culture

Anther culture means growing of pollens in a nutrient medium under aseptic conditions. Pollens or male gametes of a flower are stored inside a pollen sac. Pollens are haploids and contain a half of the total chromosomes present in a mother plant. Because of that, pollen culture gives opportunity of producing a large number of haploid plants within a short time period.

Anther culture can be performed in two main ways;

1. Direct embryogenesis

    ↳ Production of plants by allowing embryogenesis from pollens (Figure 1.7).

2. Indirect embryogenesis

> Production of callus from pollens and allow callus to differentiate into complete plants. This method allows easy production of plantlets, which has become more popular.

Culturing of paddy pollens is a best example for a crop using anther culture for plant propagation. Other than that anther culture is used over 200 species of plants including barley, tobacco and many more.

This technique is very much important for the production of large number of haploids with least time and is very easy in some species that induces cell dissection in immature pollen compared to other methods.

Figure 1.7 Callus induction from anthers, regeneration, rooting and hardening of rice. Callus induction (a), Callus proliferation (b), Initiation of green shoots(c), Initiation of albino shoots (d), Elongation of green shoots(e), Elongation of albino shoots (f), Initiation of roots in rooting media (g), Hardening and acclimitazation of green plants (h)
Source : http://scialert.net/fulltext/?doi=ajbkr.2011.470.477

Callus culture

Culturing of live plant parts in artificial nutrient medium under aseptic conditions to produce callus is known as callus culture. Callus is an undifferentiated mass of cells to perform any function. But, all these calli do not have the ability to differentiate into plants. It is very much important to identify the calli which do not possess the ability to develop into complete plants. A callus with a generating ability shows a dryness on its surface and shows a somewhat dispersed nature. After a correct identification of calli with the ability to regenerate into complete plants should be introduced to nutrient media with different hormones for their further development.

The callus culture is often sustained on a gel medium, which is composed of agar and a mixture of given macro and micronutrients depending on the type of cells. Different types of basal salt mixtures such as Murashige and Skoog (MS) medium is also used in addition to vitamins to enhance growth. After introduction of explant parts into the media, they should be placed in dark to induce callusing for many plant species. For this culture vessels can be covered using a black polythene and placed in the culture room. But, some tissue culture laboratories consist of a small dark place within the culture area. One of the best examples for callus culture is production of Anthurium plantlets using leaf explants.

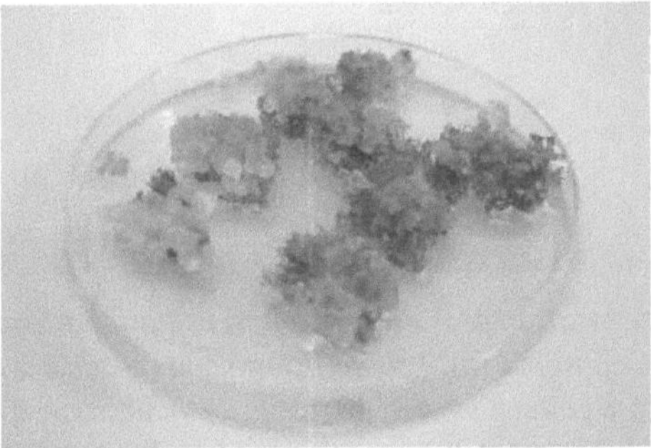

Figure 1.8 Callus growing on a nutrient medium

Figure 1.9 In vitro regeneration of A. andraeanum using foliar explants; Callus initiation from pale green leaf segments (a), Proliferating callus after 8 weeks of subculture (b), Shoot regeneration from callus tissue in 10 weeks of incubation (c), Rooted plantlets (d), Plantlets establishing in the green house (e), Hardened anthurium plants (f)

2. History of Plant Tissue Culture

Year	Worker	Contribution
1902	C.Haberlant	First attempt to culture isolated plant cells in vitro on artificial medium
1922	WJ Robbins and W. Kotte	Culture of isolated roots (for short periods) (organ culture)
1934	P R White	Demonstration of indefinite culture of tomato roots (long period)
1939	R J Gautheret and P Nobecourt	First long term plant tissue culture of callus, involving explants of cambail tissues isolated from carrot.
1939	P R White	Callus culture of tobacco tumor tissues from intersepcific hybird of *Nicotina glaucum* X *N.longsdorffi*
1941	J Van Overbeek	Discovery of nutritional value of liquid endosperm of coconut for culture of isolated carrot embryo.
1942	P R White and A C Braun	Experiments on crownn-gall and tumor formation in plants, growth of bacteria free crown-gall tissues.
1948	A Caplan and F C Stewart	Use of coconut milk plus 2, 4-D fro proliferation of cultured carrot and potato tissues
1950	G Morel	Culture of monocot tissues using coconut milk.
1953	W H Muir	Inoculation of callus pieces in liquid medium can give a suspension of single cells amenable tosubculture. Development of technique for culture of single isolated cells.

1953	W Tulecke	Haploid culture from pollen of gymnosperm (Ginkgo)
1955	C O Miller, F Skeog and others	Discovery of cytokinins. E.g. Kinetin, or potent cell division factor.
1955	E ball	Culture of gymnosperm tissues (Sequoia)
1957	F Skoog and C O Miller	Hypotheses that shoot and root initiation in cultured callus is regulated by the proportion of auxins and cytokinins in the culture medium.
1960	E C Cocking	Enzymatic isolation and culture of protoplast.
1960	G Morel	Development of shoot apex culture technique.
1964	G Morel	Use of modified shoot apex technique for orchid proportion.
1966	S G Guha and S C Maheshwari	Cultured anthers and pollen and produce haploid embryos.
1974	J P Nitsch	Culture of microspores of Datura and Nicotina, to double the chromosome number and to harvest seed from homozygous diploid plants just within five months.
1978	G Melchers	Production of somatic hybrids from attached to plasmid vectors into naked plant protoplast.
1983	K A Barton , W J Brill and J H Dodds Bengochea	Insertion of foreign genes attached to plasmid vectors into naked plant protoplast.
1983	M D Chilton	Production of transformed tobacco plants following single cell transformation or gene insertion.

3. Organization of a Tissue Culture Laboratory

Following factors should be considered prior to the place selection to build a tissue culture laboratory;

1. Availability of electricity
2. Availability of transport facility
3. Ability to maintain the cleanliness around the building
4. Easiness to obtain necessary raw materials for culturing

Basic Equipment for Tissue Culture Laboratory

The equipment mentioned below are essential to maintain a tissue culture laboratory.

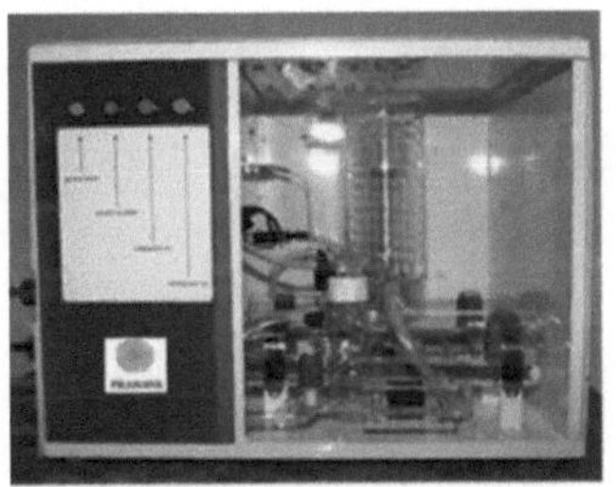

Water Distillation Unit

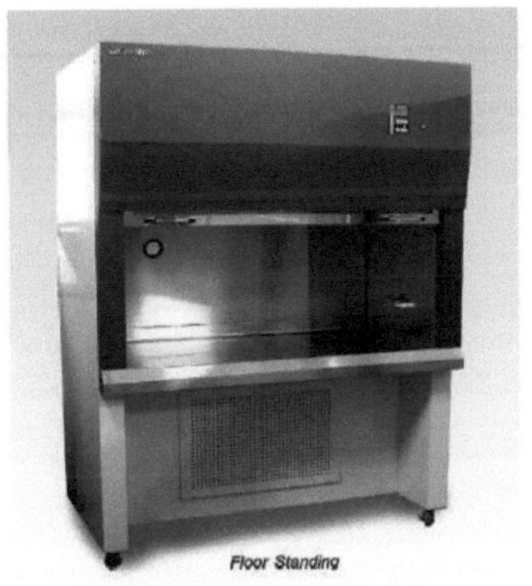

Laminar Air Flow Cabinet

Electronic Balance

Magnetic Stirrer and hot plate

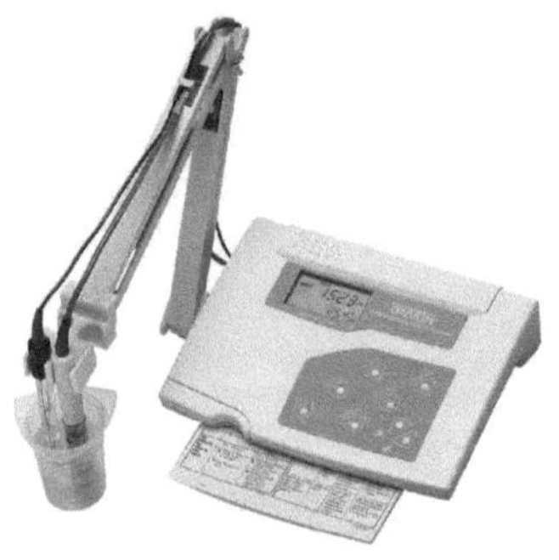

pH meter and EC meter

Sealer

Auto clave

Refrigerator

Microwave oven

Drying Oven

Table 3.1 Uses of basic equipment used in plant tissue culture laboratory

Equipment	Function
Water distillation unit	Production of distilled water to prepare media
Autoclave	Sterilization of media, glassware and tools used for culturing
pH meter	Measures pH of media during preparation
Magnetic stirrer	Proper mixing chemicals with distilled water during media preparation
Refrigerator	Store chemicals and hormones which degrade at room temperature which are necessary for media preparation
Laminar air flow cabinet	Supply of aseptic conditions during culturing process
Electronic balance	Measure chemicals
Electric oven	Sterilize glassware
Sealer	Seal polythene
Microwave oven	Dissolve agar, necessary for gelling media

Basic Sections of a Tissue Culture Laboratory

Presence of following sections in the tissue culture laboratory helps in making the tasks easily managed and organized. It is better to arrange these sections as in the given diagram.

- Washing area
- Media preparation and sterilization area
- Culture initiation area
- Culture growing area
- Office

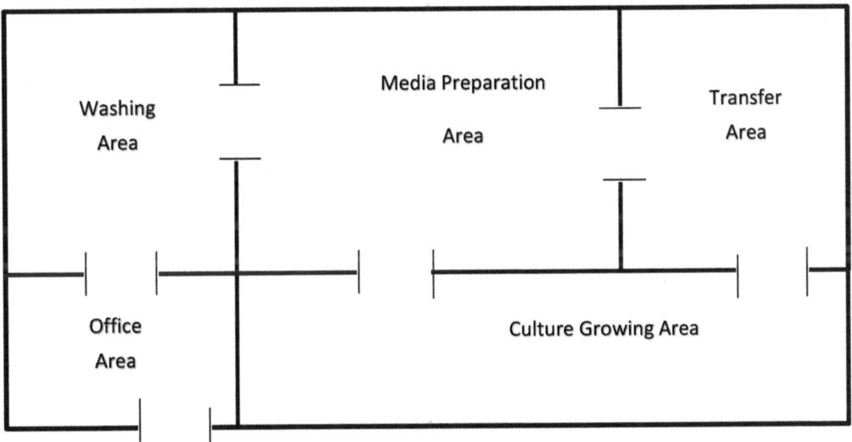

Figure 3.1 Design of a general tissue culture laboratory

Washing area

The washing area should consist of main components including taps, drainage basin, drainage racks and a covered area to store cleaned vessels from dust. It is better to construct the washing area near to a window which receives maximum amount of sunlight to the area.

Washing area should contain large sinks, some lead-lined to resist acids and alkalis, draining boards, and racks, and have access to distilled water. Space for drying ovens or racks, automated dishwashers, acid baths, pipette washers and driers, and storage cabinets should also be available in the washing area.

Media preparation and sterilization area

The media preparation area should have plenty of storage space for the chemicals, culture vessels and glassware required for media preparation and dispensing. Bench space for hot plates/stirrers, pH meters, balances, water baths, and media dispensing equipment should be available. Other necessary equipment may include distilled water and refrigerators for storing stock solutions and chemicals, a microwave or a convection oven, and an autoclave for sterilizing media, glassware, and instruments.

In preparing culture media, analytical grade chemicals should be used and good weighing habits practiced. To secure accuracy, step-by-step routine should be developed for media preparation. The water used in preparing media must be pure with highest quality.

Transfer Area

Under very clean and dry conditions, tissue culture techniques can be successfully performed on an open laboratory bench. However, it is advisable that a laminar flow hood or sterile transfer room be utilized for making transfers. Within the transfer area there should be a source of electricity, gas, compressed air and vacuum. The most desirable arrangement is a small dust-free room equipped with an overhead ultraviolet light and a positive pressure ventilation unit. The ventilation should be equipped with a high-efficiency particulate air (HEPA) filter. A 0.3-µm HEPA filter of 99.97-99.99% efficiency works well. All surfaces in the room should be designed and constructed in such a manner that dust and microorganisms do not accumulate and the surfaces can be thoroughly cleaned and disinfected. A room of such design is particularly useful if large numbers of cultures are being manipulated or large pieces of equipment are being utilized. Another type of transfer area is a laminar flow hood. Air is forced into the unit through a dust filter then passed through a HEPA filter. The air is then either directed downward (vertical flow unit) or outward (horizontal flow unit) over the working surface. The constant flow of bacteria-free filtered air prevents nonfiltered air and particulate matter from settling on the working surface. The simplest type of transfer area suitable for tissue culture work is an enclosed plastic box commonly called a glove box. This type of culture hood is sterilized by an ultraviolet light and wiped down periodically with 95% ethyl alcohol when in use. This type of unit is used when relatively few transfers are required.

Culture Room

All types of tissue cultures should be incubated under conditions of well-controlled temperature, humidity, air circulation and light quality and duration. These environmental factors may influence the growth and differentiation process directly during culture or indirectly by affecting their response in subsequent generations. Protoplast cultures, low-density cell suspension cultures and anther cultures are particularly sensitive to environmental cultural condition. Typically, the culture room for growth of plant tissue cultures should have a temperature between 15° and 30°C, with a temperature fluctuation of less than ±0.5°C; however, a wider range in temperature may be required for specific experiments. It is also recommended that the room have an alarm system to indicate when the temperature has reached preset high or low temperature limits, as well as continuous temperature recorder to monitor temperature fluctuations. The temperature should be constant throughout the entire culture room (i.e., no hot or cold spots). The culture room should have enough fluorescent lighting to reach the range 5000 - 10,000 lux; the lighting should be adjustable in terms of quantity and photoperiod duration. Both light and temperature should be controllable for a 24-hr period. The culture room should have fairly uniform forced-air ventilation, and a humidity range of 60-70%. Many incubators, large growth chambers, and walk-in environmental chambers meet these specifications.

Figure 3.2 Washing area

Figure 3.4 Media preparation area

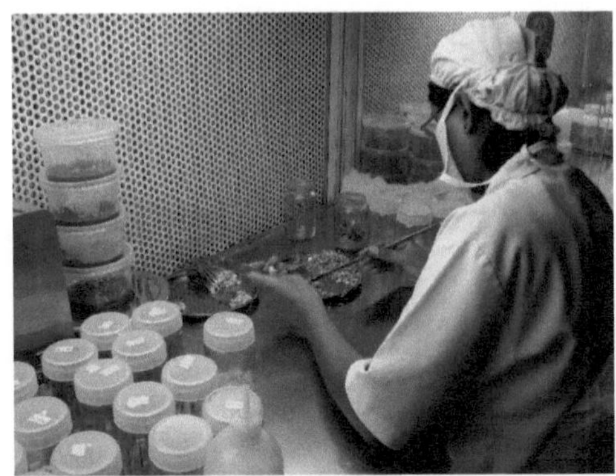

Figure 3.5 Transfer area – work on the laminar air flow cabinet

Figure 3.6 Culture Room with racks to keep culture vessels.

4. Preparation of Nutrient Media

The nutrient medium used for tissue culture should composed of following components,

- Inorganic nutrients
- Carbon energy source
- Organic substances
- Plant hormones
- Solidifying agents
- Other supplements

Inorganic nutrients

Inorganic nutrients added to media are of two types as macro nutrients and micro nutrients.

Macro nutrients

Nitrogen (N), Phosphorus (P), Potassium (K), Calcium (Ca), Magnesium (Mg) and Sulphur (S) are categorized under macro nutrients.

Micro nutrients

Iron (Fe), Manganese (Mn), Zinc (Zn), Boron (B), Copper (Cu), Molybdenum (Mo) and Cobalt (co) fall into micro nutrients required for plant growth.

Carbon energy source

At the initial stage   explant parts used for tissue culture do not have the possibility to produce energy through photosynthesis. Therefore, it is necessary to add a carbon energy source. For this purpose, sucrose, glucose, fructose and maltose can be used.

Organic substances

Vitamins and amino acids are added as organic substances to the nutrient medium. Vitamins catalyze various metabolic reactions and amino acids are used to promote cell growth. Nicotinic acid (vitamin B_3) and Pyridoxine hydrochloride (vitamin B_6) are commonly used when preparing Murishige and Skoog medium. Biotin (vitamin H), Ascorbic acid (vitamin C), thiamin hydrochloride (vitamin B_1) and Pantothenic acid (vitamin B_5) are used when preparing media.

Casein, Glutamine, Adenine and Glycine are widely used amino acids in tissue culture for media preparation. Other than the, Arginine, Cysteine and Asparanin are also used in media preparation.

Plant hormones

Plant growth regulators are important in plant tissue culture since they play vital roles in stem elongation, tropism, and apical dominance. They are generally classified into the following groups; auxins, cytokinins, gibberellins, abscisic acid and ethylene.

There are 5 main classes of plant growth regulators.

- Auxins: promote both cell division and cell growth.
- Cytokinins: promote cell division, growth and development
- Gibberellins: Major action of gibberillin is stimulation of cell stem elongation and flowering.
- Abscissic Acid: Primarily involved in water-stem responses, seed germination and enhances somatic embryogenesis.
- Ethylene: is gaseous and is associated with fruit ripening in climacteric fruits

Moreover, proportion of auxins to cytokinins determines the type and extent of organogenesis in plant cell cultures. There are few main types of plant hormones used for tissue culturing. Among them Auxin and Cytokinin are the prominent two types of hormones used widely in media preparation. Auxin is very much important for rooting, shoot elongation, totipotency and apical dominance.

    Ex: Indole Acetic Acid (IAA)

            Indole Butyric Acid (IBA)

Cytokinin plays a major role in cell division, bud formation and development

Ex:	Kinetin

Benzyl Amino Purine (BAP)

Other than that, Gibelleric acid and Absisic acid are used in small quantities in tissue culture techniques.

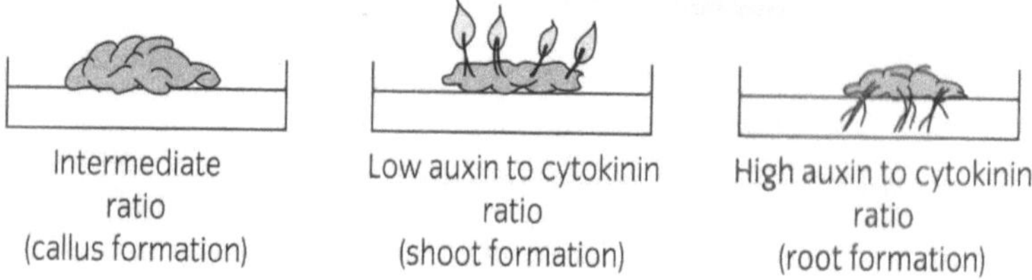

Figure 4.1 Effects of different ratios of plant growth substances on tissue culture

Solidifying agents

Solidifying agents are used to prepare semi solid and solid culture media. Agar is the most widely used solidifying agent in media preparation. This agent is extracted from a seaweed, red algae. Other than that, corn flour or gelatin can be used to solidify media.

Other supplements

Some media are supplemented with natural substances or extracts such as protein hydrolysates, coconut milk, yeast extract, malt extract, ground banana, orange juice and tomato juice, to test their effect on growth enhancement. A wide variety of organic extracts are now commonly added to culture media. The addition of activated charcoal is sometime added to culture media where it may have either a beneficial or deleterious effect. Explanation of the mode of action of activated charcoal was based on adsorption of inhibitory compounds from the medium, adsorption of growth regulators from the culture medium or darkening of the medium.

MS Medium and Composition

Table 4.1 Composition of MS medium Constituent Final concentration (mg/l)

Major salt	Amount (mg)
NH_4NO_3	1650
KNO_3	1900.00
$CaCl_2.7H_2O$	440.00
$MgSO_4.7H2O$	370.00
KH_2PO_4	170.00
Minor	
KI	0.83
H_3BO_3	6.20
$MnSO_47H_2O$	22.30
$ZnSO_4$	8.6
$NaMoO_4$	0.25
$CuSO_4$	0.025
$CoCl_2$	0.025
Iron Source	
$FeSO_4.7H_2O$	27.8
$Na-EDTA\ 2H_2O$	37.3
Sucrose	30000

Vitamins and organics

- Myo-Inositol 100 mg/l
- Nicotinic Acid 0.5 mg/l
- Pyridoxine · HCl 0.5 mg/l
- Thiamine · HCl 0.1 mg/l

- Glycine 2 mg/l
- Lactalbumin Hydrolysate (Edamin) (optional) 1 g/l

Murashige and Skoog medium (MS) is a plant growth medium used in the laboratories for plant tissue culture. MS medium was invented by plant scientists called, Toshio Murashige and Folke K. Skoog in 1962 during Murashige's search for a new plant growth regulator. Along with its modifications, it is the most commonly used medium in plant tissue culture experiments in laboratories.

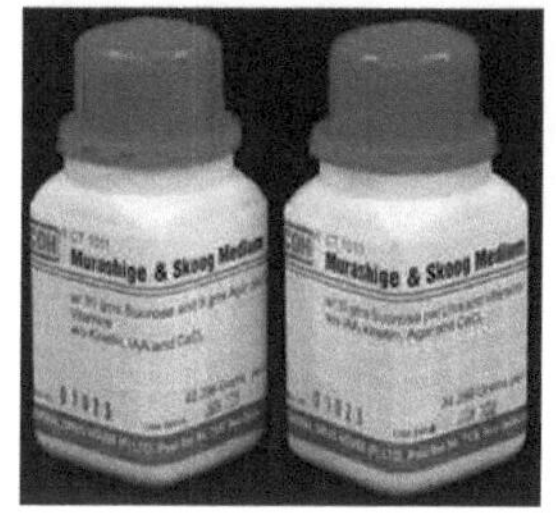

Murashige and Skoog medium is the first choice as its having balanced composition over other media. The composition of macro, micro nutrients, vitamins and organics are highly suitable for majority of the plant species. However, for specific plant species and based on type of ex-plant (meristem /leaf/petal/anther/ovule/embryo/seed) sometimes it is necessary use other type of media or another modified version of MS media.

Table 4.2 Comparison of Common Plant Tissue Culture Media

Constituent	Concentration in culture medium (mg/liter)		
	MS	SH (Shenck and Hildebrandt)	B5
KNO_3	1900	2500	2500
NH_4NO_3	1650		
$NH_4H_2PO_4$		300	
$(NH_4)_2SO_4$			134
$MgSO_4 \cdot 7H_2O$	370	400	250
$CaCl_2 \cdot 2H_2O$	440	200	150
KH_2PO_4	170		
$NaH_2PO_4 \cdot H_2O$			150
$MnSO_4 \cdot H_2O$		10.0	10.0
$MnSO_4 \cdot 4H_2O$	22.3		
KI	0.83	1.0	0.75
H_3BO_3	6.2	5.0	3.0
$ZnSO_4 \cdot 7H_2O$	8.6	1.0	2.0
$CuSO_4 \cdot 5H_2O$	0.025	0.2	0.025
$Na_2MoO_4 \cdot 2H_2O$	0.25	0.1	0.25
$CoCl_2 \cdot 6H_2O$	0.025	0.1	0.025
$FeSO_4 \cdot 7H_2O$	27.8	15.0	27.8
Na_2EDTA	37.3	20.0	37.3

Nicotinic acid	0.5	5.0	1.0
Pyridoxine-HCl	0.5	0.5	1.0
Thiamine-HCl	0.1	5.0	10.0
myo-Inositol	100	1000	100
Glycine	2.0		
Sucrose	30000	30000	20000

In complex media such as these, consisting of many components, there are a huge number of permutations of substances and concentrations to test to compose an ideal medium for a particular plant species and genotype. When working with a species new to you, the first place to start is in the literature, finding out what other people have used for that species or one closely related. However, there may be considerable differences in requirements for different cultivars. In practice, researchers may test several basal media, but only try to optimize a few components, in particular, plant growth regulators (PGRs).

5. Micropropagation

Micropropagation is the aseptic culture of cells, pieces of tissue, or organs using modern plant tissue culture methods. It is possible to regenerate new plants from small pieces of plant tissue because each cell of a given plant has the same genetic makeup and is totipotent, that is, capable of developing along a "programmed" pathway leading to the formation of an entire plant that is identical to the plant from which it was derived.

In addition to its biotechnological applications, micropropagation is used commercially to asexually propagate plants. Using micropropagation, millions of new plants can be derived from a single plant. This rapid multiplication allows breeders and growers to introduce new cultivars much earlier than they could by using conventional propagation techniques, such as cuttings. Micropropagation also can be used to establish and maintain virus-free plant stock. This is done by culturing the plant's apical meristem, which typically is not virus-infected, even though the remainder of the plant may be. Once new plants are developed from the apical meristem, they can be maintained and sold as virus-free plants. Micropropagation differs from all other conventional propagation methods in that aseptic conditions are essential to achieve success. The process of micropropagation can be divided into five stages:

Stage 0

The initial step of micro-propagation in which stock plants has to be grown under controlled condition to obtain explants before using for culture initiation. These should be free from diseases and pests. For micropropagation, stock plants at vegetative stage is better, but if plants are at mature stage, remove the apical bud and spray growth hormones to induce vegetative growth. Stock plants should be kept inside enclosed houses, and should apply insect repellents and required fungicides.

Factors to be considered when selecting mother plants:

- Age of the plant
- Season
- Explant size
- Plant quality

Stage I

A piece of plant tissue (called an explant) is,

(a) seperated from the plant, (b) disinfested (removal of surface contaminants), and (c) placed on a medium. A medium typically contains mineral salts, sucrose, and a solidifying agent such as agar. The objective of this stage is to achieve an aseptic culture. An aseptic culture is one without contaminating bacteria or fungi.

Stage II

A growing explant can be induced to produce vegetative shoots by including a cytokinin in the medium. A cytokinin is a plant growth regulator that promotes shoot formation from growing plant cells.

Stage III

Growing shoots can be induced to produce adventitious roots by including an auxin in the medium. Auxins are plant growth regulators that promote root formation. For easily rooted plants, an auxin is usually not necessary and many commercial labs will skip this step.

Stage IV

A growing, rooted shoot can be removed from tissue culture and placed in soil. When this is done, the humidity must be gradually reduced over time because tissue-cultured plants are extremely susceptible to wilting. The optimal temperature for culture is in the range of 20-28°C (for majority 24-26°C). Lower light intensity is more appropriate for good micro propagation.

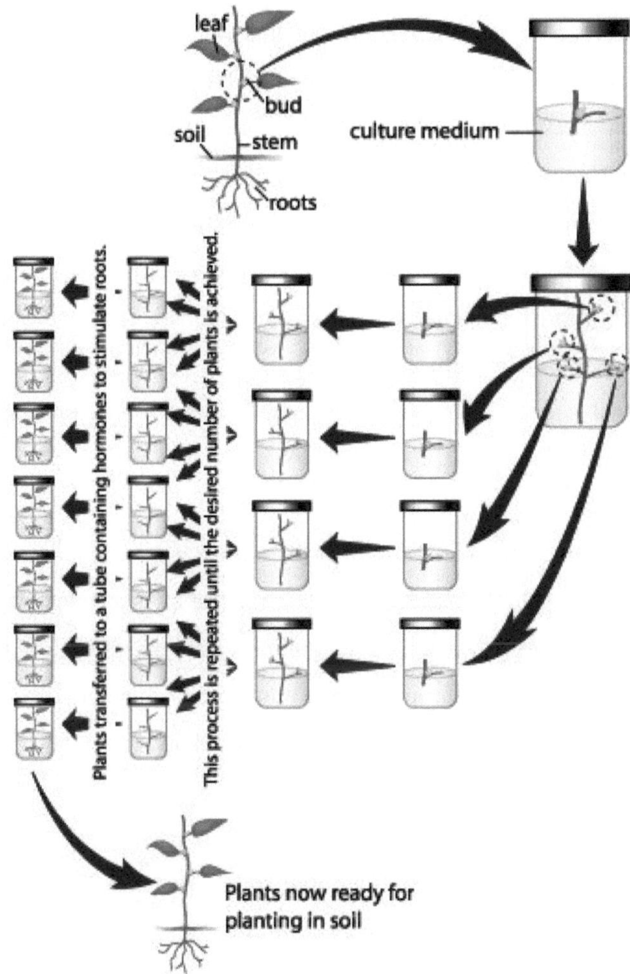

Figure 5.1 Steps of micropropagation

Methods of Micropropagation

Micro propagation mostly involves *in vitro* clonal propagation by two approaches:

1. Multiplication by axillary buds/apical shoots

2. Multiplication by adventitious shoots

Besides the above two approaches, the plant regeneration processes namely organogenesis and somatic embryogenesis may also be treated as micropropagation.

3. Organogenesis: The formation of individual organs such as shoots, roots, directly from an explant (lacking preformed meristem) or from the callus and cell culture induced from the explant.

4. Somatic embryogenesis: The regeneration of embryos from somatic cells, tissues or organs.

1. Multiplication by Axillary Buds and Apical Shoots:

Actively dividing meristems are present at the axillary and apical shoots (shoot tips). The axillary buds located in the axils of leaves are capable of developing into shoots. In the in vivo state, however only a limited number of axillary meristems can form shoots. By means of induced in vitro multiplication in micro propagation, it is possible to develop plants from meristem and shoot tip cultures and from bud cultures.

<u>Meristem and Shoot Tip Cultures:</u>

Apical meristem is a dome of tissues located at the tip of a shoot. The apical meristem along with the young leaf primordia constitutes the shoot apex. For the development of disease-free plants, meristem tips should be cultured. Meristem or shoot tip is isolated from a stem by a V-shaped cut. The size (frequently 0.2 to 0.5 mm) of the tip is critical for culture. In general, the larger the explant (shoot tip), better are the chances for culture survival. For good results of micro propagation, explants should be taken from the actively growing shoot tips, and the ideal timing is at the end of the plants dormancy period.

One of the most widely used medium for meristem culture is MS medium. A diagrammatic representation of shoot tip (or meristem) culture in micro propagation is given in Figure 5.2, and briefly described here under.

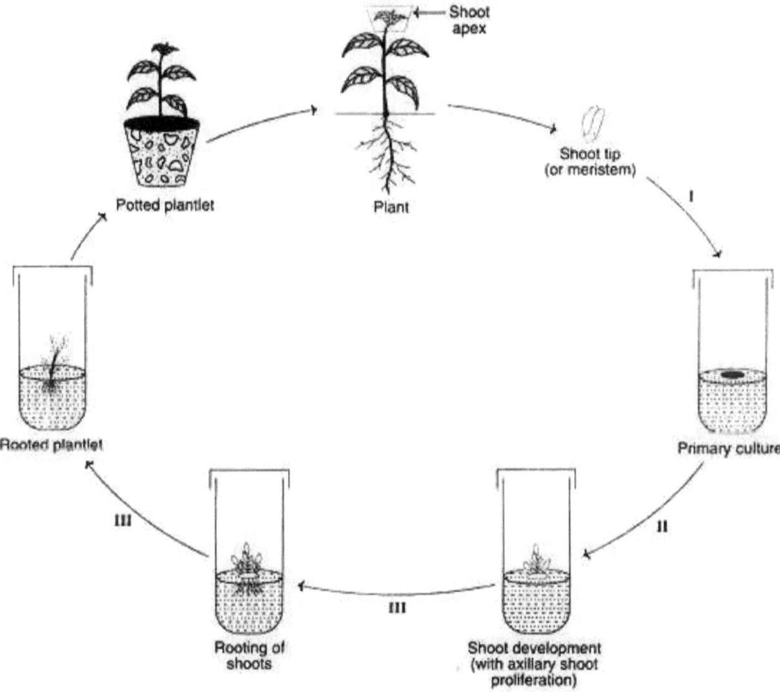

Figure 5.2 Steps of shoot tip (or meristem) culture in micro propagation. I, II and III represents stages in micro propagation respectively.

<u>Bud Cultures</u>:

The plant buds possess quiescent meristems depending on the physiological state of the plant. Two types of bud cultures are used;

- Single node culture
- Axillary bud culture.

Single node culture:

A bud along with a piece of stem is isolated and cultured to develop into a plantlet. Closed buds are used to reduce the chances of infections. In single node culture, no cytokinin is added.

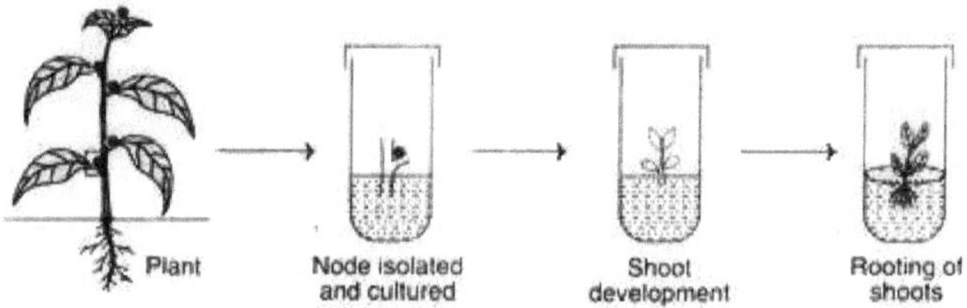

Figure 5.3 Steps of single node culture

Axillary bud culture:

In this method, a shoot tip along with axillary bud is isolated. The cultures are carried out with high cytokinin concentration. To induce axillary bud formation by inhibiting apical dominance.

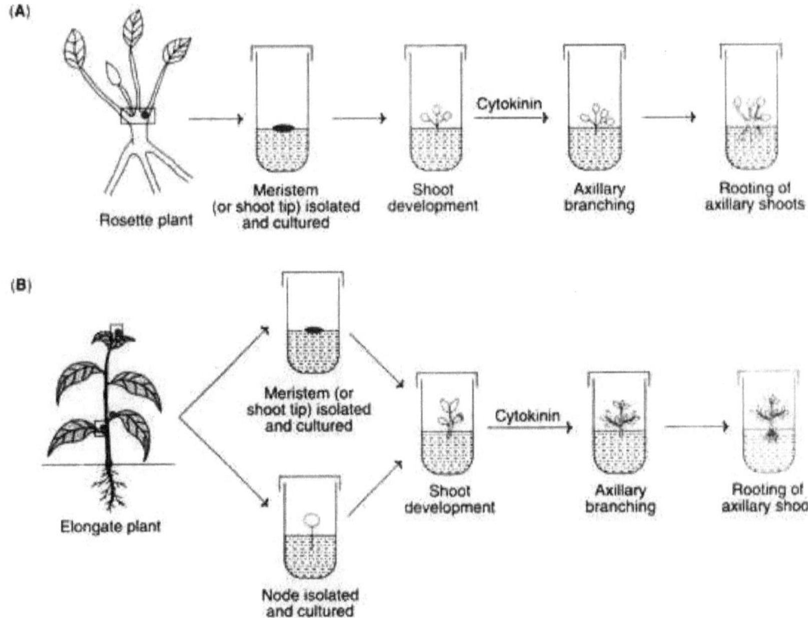

Figure 5.4 Steps of micro propagation of plants by axillary bud (or shoot tip) (A) Rosette plant(bush), (B) Elongate plant

2. Multiplication by Adventitious Shoots:

The stem and leaf structures that are naturally formed on plant tissues located in sites other than the normal leaf axil regions are regarded as adventitious shoots. There are many adventitious shoots which include stems, bulbs, tubers and rhizomes. The adventitious shoots are useful for *in vivo* and *in vitro* clonal propagation. The meristematic regions of adventitious shoots can be induced in a suitable medium to regenerate to plants.

3. *Organogenesis:*

Organogenesis is the process of morphogenesis involving the formation of various plant organs such as shoots, roots, flowers, buds from explant or cultured plant tissues. It is of two types;

1. direct organogenesis

2. indirect organogenesis

<u>Direct organogenesis</u>:

Tissues from leaves, stems, roots and inflorescences can be directly cultured to produce plant organs. In direct organogenesis, the tissue undergoes morphogenesis without going through a callus or suspension cell culture stage. The term direct adventitious organ formation is also used for direct organogenesis.

Induction of adventitious shoot formation directly on roots, leaves and various other organs of intact plants is a widely used method for plant propagation. This approach is particularly useful for herbaceous species. For appropriate organogenesis in culture system, control of the levels of growth regulators, auxin and cytokinin is required.

<u>Indirect organogenesis</u>:

When the organogenesis occurs through callus or suspension cell culture formation is known as indirect organogenesis. Callus growth can be established from many explants (leaves, roots, cotyledons, stems, flower petals etc.) for subsequent organogenesis.

It is highly advantageous to select meristematic tissues (shoot tip, leaf, and petiole) for efficient indirect organogenesis. This is because their growth rate and survival rate are much better.

For indirect organogenesis, the cultures may be grown in liquid medium or solid medium. Many culture media (MS, B5 White's etc.) can be used in organogenesis. The concentration of growth regulators in the medium is critical for organogenesis.

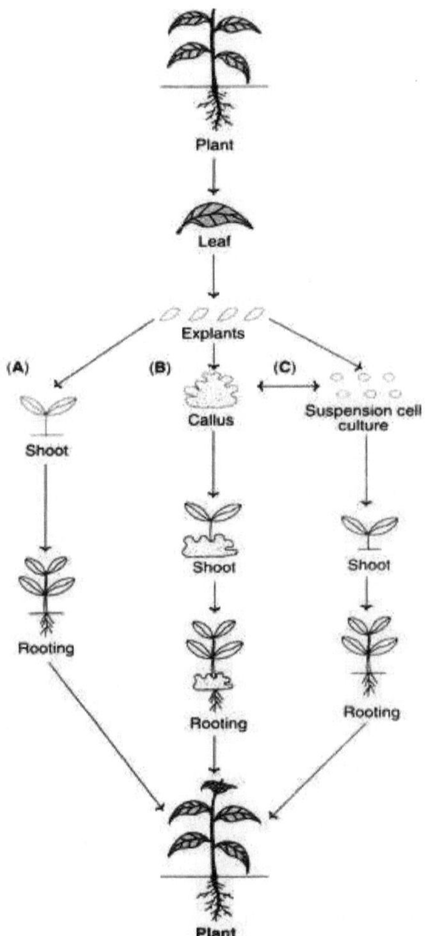

oioure 5.5 Micropropaoation of plants by oroanooenesis. (A) Direct oroanooenesis (B) Indirect oroanooenesis throuoh callus (C) Indirect oroanooenesis throuoh suspension culture

By varying the concentrations of auxins and cytokinins, *in vitro* organogenesis can be manipulated:

i. Low auxin and low cytokinin concentration will induce callus formation.

ii. Low auxin and high cytokinin concentration will promote shoot organogenesis from callus.

iii. High auxin and low cytokinin concentration will induce root formation.

4. Somatic Embryogenesis:

The process of regeneration of embryos from somatic cells, tissues or organs is regarded as somatic (or asexual) embryogenesis. Somatic embryogenesis may result in non-zygotic embryos or somatic embryos (directly formed from somatic organs), parthogenetic embryos (formed from unfertilized egg) and androgenic embryos (formed from male gametophyte).

In a general usage, when the term somatic embryo is used it implies that it is formed from somatic tissues under *in vitro* conditions. Somatic embryos are structurally similar to zygotic (sexually formed) embryos, and they can be excised from the parent tissues and induced to germinate in tissue culture media.

Development of somatic embryos can be done in plant cultures using somatic cells, particularly epidermis, parenchymatous cells of petioles or secondary root phloem. Somatic embryos arise from single cells located within the clusters of meristematic cells in the callus or cell suspension. First a pro-embryo is formed which then develops into an embryo, and finally a plant.

Two routes of somatic embryogenesis are known — direct and indirect (Figure 5.6).

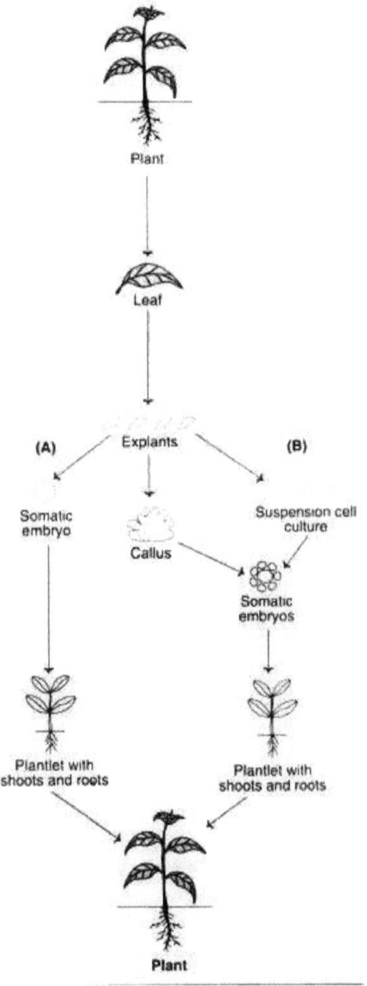

Figure 5.7 Micropropagation of plants by embryogenesis. (A) Direct embryogenesis (B) Indirect embryogenesis

<u>Direct Somatic Embryogenesis</u>:

When the somatic embryos develop directly on the excised plant (explant) without undergoing callus formation, it is referred to as direct somatic embryogenesis (Figure 5.7A). This is possible due to the presence of pre-embryonic determined cells found in certain tissues of plants. The characteristic features of direct somatic embryogenesis are to avoid the possibility of introducing somaclonal variations in the propagated plants.

<u>Indirect Somatic Embryogenesis</u>:

In indirect embryogenesis, the cells from explant (excised plant tissues) are made to proliferate and form callus, from which cell suspension cultures can be raised. Certain cells referred to as induced embryo genic determined cells from the cell suspension can form somatic embryos. Embryogenesis is made possible by the presence of growth regulators (in appropriate concentration) and under suitable environmental conditions. Somatic embryogenesis (direct or indirect) can be carried on a wide range of media (e.g. MS, White's). The addition of the amino acid L-glutamine promotes embryogenesis. The presence of auxin such as 2, 4-dichlorophenoxy acetic acid is essential for embryo initiation. On a low auxin or no auxin medium, the embryogenic clumps develop into mature embryos.

Indirect somatic embryogenesis is commercially very attractive since a large number of embryos can be generated in a small volume of culture medium. The somatic embryos so formed are synchronous and with good regeneration capability.

Artificial Seeds from Somatic Embryos

Artificial seeds can be made by encapsulation of somatic embryos. The embryos, coated with sodium alginate and nutrient solution, are dipped in calcium chloride solution. The calcium ions induce rapid cross-linking of sodium alginate to produce small gel beads, each containing an encapsulated embryo. These artificial seeds (encapsulated embryos) can be maintained in a viable state till they are planted.

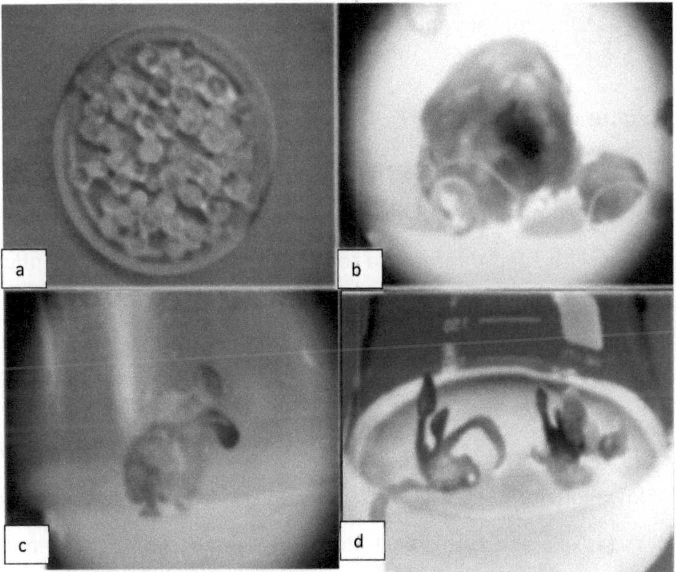

Figure 5.8 Encapsulated somatic embryo (a), initial stage of seed germination (b) germinated seed (c), plant obtained from synthetic seed and (d)

Factors Affecting Micropropagation

For a successful in vitro clonal propagation (micro propagation), optimization of several factors is needed. Some of these factors are briefly described.

1. Genotype of the plant:

Selection of the right genotype of the plant species (by screening) is necessary for improved micro propagation. In general, plants with vigorous germination and branching capacity are more suitable for micropropagation.

2. Physiological status of the explants:

Explants (plant materials) from more recently produced parts of plants are more effective than those from older regions. Good knowledge of donor plants' natural propagation process with special reference to growth stage and seasonal influence will be useful in selecting explants.

3. Culture media:

The standard plant tissue culture media are suitable for micro propagation during stage I and stage II. However, for stage III, certain modifications are required. Addition of growth regulators (auxins and cytokinins) and alterations in mineral composition are required. This is largely dependent on the type of culture (meristem, bud etc.).

4. Culture environment:

<u>Light:</u>

Photosynthetic pigment in cultured tissues absorb light and influence micropropagation. The quality of light is important to influence *in vitro* growth of shoots, e.g. blue light induced bud formation in tobacco shoots. Variations in diurnal illumination also influence micro propagation. In general, an illumination of 16 hours day and 8 hours night is satisfactory for shoot proliferation.

<u>Temperature:</u>

For micropropagation the optimal temperature is about 25°C. But, there may be some exceptions according to the plant variety.

<u>Composition of gas phase:</u>

The constitution of the gas phase in the culture vessels also influences micro propagation. Unorganized growth of cells is generally promoted by ethylene, O_2, CO_2, ethanol and acetaldehyde.

<u>Factors affecting *In vitro* rooting:</u>

For efficient *in vitro* rooting during micropropagation, low concentration of salts (reduction to half to one quarter from the original) is advantageous. Induction of roots is also promoted by the presence of suitable auxin (NAA or IBA).

6. Initiation and Aseptic Culture Establishment

The most necessary fundamental is to maintenance of aseptic conditions throughout the whole process.

I. Instruments and chemicals for media preparation:

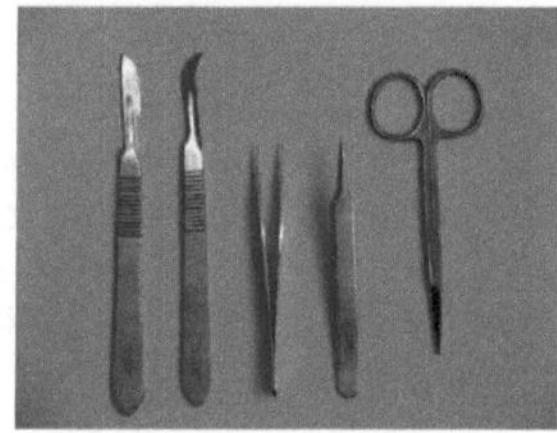 

II. Properly planned tissue culture laboratory

III. Operator and uniform

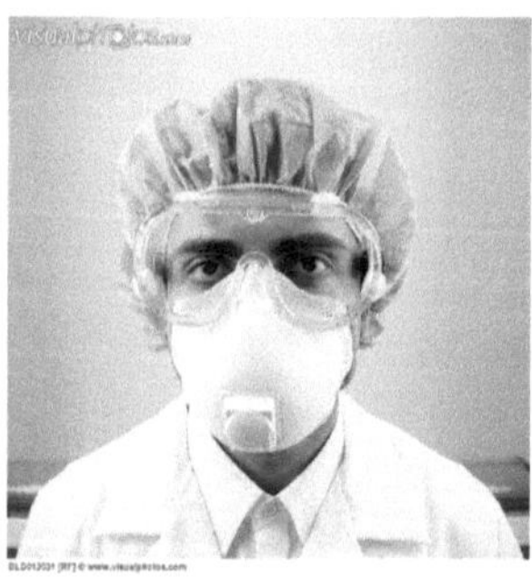 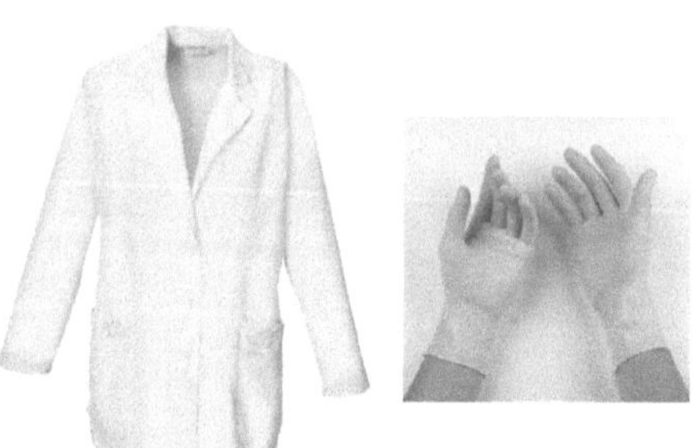

IV. Explants

V. Sterilizing agents

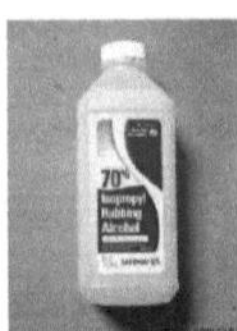

- Liquid laundry bleach (NaOCl at 5-6% by vol)

    1. Rinse thoroughly after treatment

    2. Usually diluted 5-20% v/v in water; 10% is most common

- Calcium hypochlorite – $Ca(OCl)_2$

- Ethanol (EtOH)

    1. 95% used for disinfesting plant tissues

    2. Kills by dehydration

    3. Usually used at short time intervals (10 sec – 1 min)

    4. 70% used to disinfest work surfaces, worker hands

- Isopropyl alcohol (rubbing alcohol) is sometimes recommended

7. Merits and Demerits in Micropropagation

Merits of Micropropagation

Micropropagation offers several advantages over conventional propagation methods (Vegetative and sexual) through large scale production

- Shoot multiplication can be achieved in small space – because miniature plantlets are produced
- Only a small amount of initial tissue is needed for large number of clonal plants
- No seasonal fluctuations, Continuous propagation year round
- Homogenous plants
- Plantlets produced are free from micro organisms (Elimination of entophytic diseases)
- Propagation is carried out under sterile conditions – No damage is caused due to insects and diseases
- In case of virus- free material is used, a large number of virus– free plants can be obtained
- No special care is required between two subcultures as compared to conventional vegetative systems (budding/grafting)
- Mother plant or genotype of stock plant can be stored and maintained in vitro without damaging from environmental factors
- The plants, which are difficult to propagate vegetatively by conventional methods, can be propagated by this method
- Specialized techniques for growth control (micro grafting onto dwarfing rootstocks)
- Being sterile, transport across countries is permissible without difficulties

Demerits of micro propagation

- Capital intensive

  Micropropagation methods through TC involve capital intensive expensive materials like autoclave, laminar air flow bench, controlled culture rooms *etc*

- Skilled type work

  This is a technically skilled work Knowledge about material, techniques and decision making are required in persons

- Contaminations

  Contamination is a serious threat and cause severe damage to material and add substantively lot to the cost of production

- Skilled type work

  This is a technically skilled work Knowledge about material, techniques and decision making are required in persons.

- Genetic stability is doubtful in certain methods

- High cost of plantlets – It is a capital-intensive industry, if plants are produced in small number they cost too much. Cost is a major factor for the production and sale of tissue culture raised plants.

- Vitrification may occur which reduces the rate of growth multiplication of the plant and eventually causes death.

8. Applications of Plant Tissue Culture

Agriculture

➢ Anther & pollen culture in the production of haploid plants.

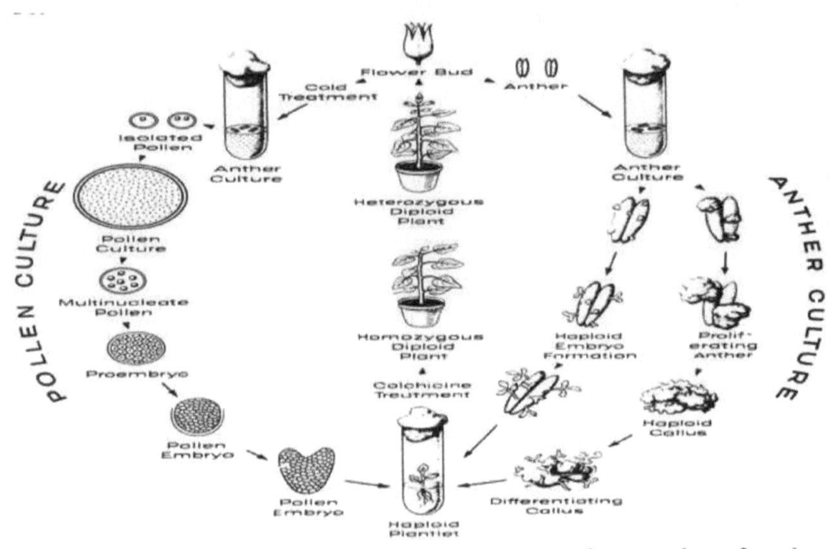

Figure 8.1 Steps of Anther and pollen culture in production of haploid plantlets

➢ Production of virus-free plants for safe germplasm transfer.

➢ Screening of *in vitro* lines for stress and disease resistance.

➢ Somaclonal Variations has been used in plant breeding programmes where the genetic variations with desired or improved characters are introduced into the plants.

Horticulture and Forestry

➢ Micropropagation method is used for rapid multiplication of ornamental plants as well as important trees yielding high fuel, pulp, fruits or oil at a large scale (Table 8.1).

   o Examples: Oil palm, *in vitro* regeneration and genetic transformation of conifers

Industries

➢ Plant cell culture is used for biotransformation (modification of functional groups of organic compounds by living cells).

➢ Food and agricultural biotechnologists are involved in using tools of molecular biology to enhance the quality and quantity of foods and economic crops. For example, Golden Rice was genetically enhanced with added beta carotene, which is a precursor to Vitamin A in the human body.

➢ Plant cells can be cultured in fermenters for the industrial production of secondary metabolites using cell culture.

Possible areas for research

➢ *In vitro* micropropagation and extraction of secondary metabolites from medicinal plants.

Table8.1 Plant Species and Secondary Metabolites Obtained from Them Using Tissue Culture Techniques

Product	Plant source	Uses
Artemisin	*Artemisia spp*	Antimalarial
Capsaicin	*Capsicum annum*	Cures Rheumatic pain
Codeine	*Papaver spp.*	Analgesic
Camptothecin	*Campatotheca accuminata*	Anticancer
Quinine	*Cinchona officinalis*	Antimalarial

9. Use of Low Cost Tissue Culture Materials for the Initiation and Multiplication of Plants

In the world over 50,000 plant varieties are successfully propagated using tissue culture and 95% of these are propagated by about 600 commercial laboratories scattered all over the developed world, but when consider about the developing countries, the tissue culture is at still developing stage.

Following limitations of conventional tissue culture have become barriers for development of tissue culture in developing countries

- High cost of equipment
- High cost of nutrient media and sterilizing agents
- High cost of facilities
- Lack or shortage of trained personnel
- Lack of systems for marketing/ delivering tissue culture products

What Low Cost Tissue Culture entails

Use of alternative and where possible locally available equipment and resources to reduce the unit cost of tissue culture products without compromising the quality of the plants.

Plant tissue culture has three components which can be intervened for cost production:

1. Chemicals (minerals nutrients, plant growth hormones, vitamins)
2. Equipment (culture containers, autoclave, laminar air flow cabinet, instruments used for micropropagation, pH meter *etc*)
3. Laboratory structures (media preparation, transfer area, culture rooms)

Points where low cost tissue culture options can be applied

1. Substitution of preparation room equipment with low cost options has been successful

   a. <u>Electric autoclave with pressure cooker</u>

   Electric autoclave is expensive; long heating time, requires electricity specialized maintenance, risk of electric shock. Pressure cooker is cheap, efficient, can use any source of heat, easy to maintain, safe.

Figure 9.1 Pressure Cooker

2. Costly, delicate pipettes and cylinders can be replaced by cheap, hardy syringes

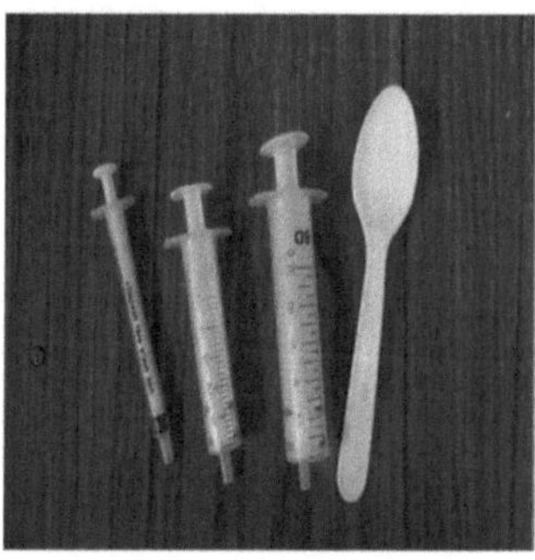

3. Instead of high cost chemical nutrients, use of low cost plant growth supporters like Maxxi crop

Low cost medium :

- Pure drinking water or rain water
- Sugar
- Albert solution
- Liquid fertilizer
- Cotton wool
- Coconut sap

Low cost medium – 1L (Ex: For banana tissue culture):

- Albert's mixture   1g
- Maxxicrop   0.2ml or 1 B vitamin tablet
- Coconut sap   50 ml
- Sugar   30g
- IAA   2ml
- BAP   5ml
- Pure drinking water or rain water to volumeup

The final pH of the medium should be around 5.6-5.8. If pH is lower than 5.6 add one drop of baking soda solution and if pH is higher than 5.8 add one drop of vinegar to adjust pH. Then, put 2g of cotton wool to the base of the culture vessel (jam bottle) and pour 10-15 ml of the medium.

Expensive agar can be replaced with Sago and some seaweed extracts.

4. Use of Vitamin B complex tablets instead of different conventional vitamins

5. NAA (Roocta) was found as effective as the conventional NAA.

6. Use of different hormone sources like king coconut water, tomato juice etc.

7. Instead of conventional electrically powered growth room use of low cost growth room under natural sun light.

Planning a low cost tissue culture laboratory

Small scale tissue culture laboratory can be planned in two ways;

1. Conversion of a prevailing room into a tissue culture laboratory (Type I)
2. Building a new tissue culture laboratory closer to the home (Type II)

Conversion of a prevailing room into a tissue culture laboratory

Select a room with 150 sqft scale. Better to select an isolated room. If it is a room from a storied a building, select a corner room at upper floor. Cleaning of the roof, walls should be easy on regular basis. If, there is no ceiling, better to use polythene to make one. The entrance should be permitted only to the workers of the laboratory.

Building a new tissue culture laboratory closer to the home

A tissue culture laboratory consists of few sections. These can be maintained as separate room or within one space (Figure 9.4). But, culture room and transfer area should be aseptic. If the laboratory is maintained in one room, separate those two areas using a plastic screen from other sections.

Different sections of the laboratory

A. Culture room

B. Transfer area

C. Media preparation section

D. Sterilizing area

E. Acclimatizing and hardening area (attached outside the room)

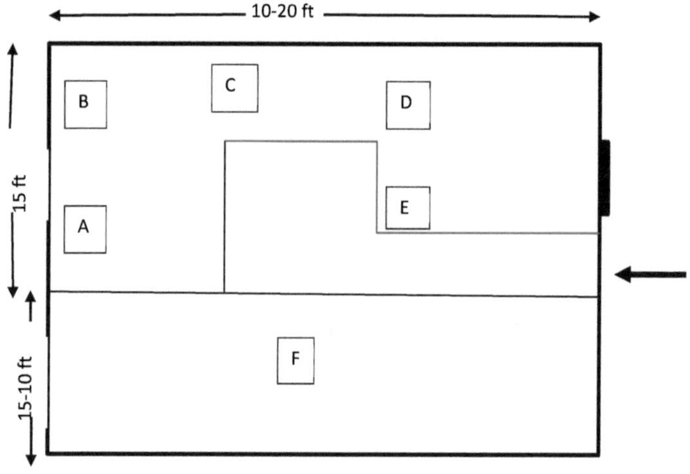

Figure 9.3 A layout of a room converted on to a tissue culture laboratory (Type I)

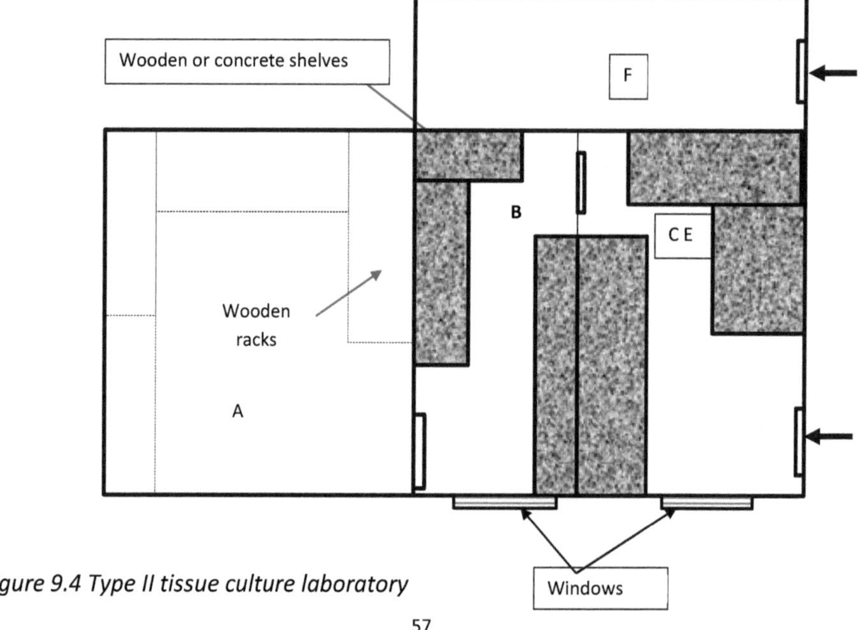

Figure 9.4 Type II tissue culture laboratory

A. Culture room

The most important section of a plant tissue culture laboratory. This area should be highly aseptic as the success of culture depends on the cleanliness of this area.

The racks to place culture vessels should be made-up of steel or wood (90 cm length X 60 cm width). 60cm long fluorescent bulbs should be set aside the racks to supply daily light requirement for 8-10 h duration. When natural light is used (Figure 9.5), instead of walls sealed glass screens should be used. If it is possible, use permeable roofing materials to the top. To drop the temperature rise, a black shade net can be used. Air conditioners are also used for this same purpose. The room temperature should be 24°C-28 °C.

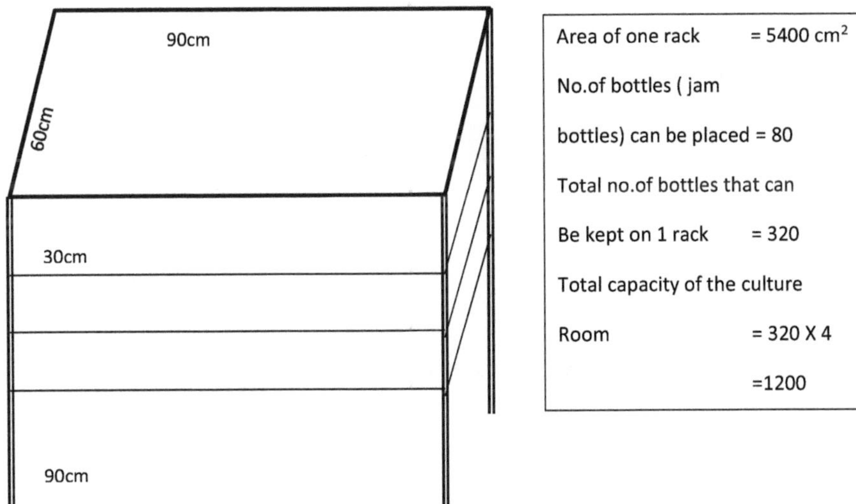

Figure 9.5 A suitable rack to place culture vessels

B. Transfer area

Instead of a laminar air flow cabinet a fish tank turned side or a chamber built by fixing glass panes to a wooden or aluminium frame can be used (Figure 9.6). This chamber should be placed in an area where minimally affected by wind or other external environmental changes far away from doors and windows.

Figure 9.6 Low cost laminar hood for plant tissue culture work

10. Addressing Problems of Plant Tissue Culture

1) Growth of bacteria and fungi in cultures

Maintenance of optimum aseptic conditions and regular inspection for contamination. As soon as a contaminated culture vessel is noted take necessary precautions to discard that culture.

2) Browning of cultures

This occurs due to presence of phenolic substances in plants. To avoid this frequent sub culturing and addition of Vitamin C (½ of a vitamin C tablet) or charcoal powder (2-5g/1L medium) can be done.

3) Yellowing and death of cultures

Reason	Solution
a) Fungi and bacteria	Destroy the cultures
b) Reduced availability of media due to growth of the cultures	Sub culturing in correct time
c) Reduced air circulation inside the bottles (due to growth of plantlets)	Sub culturing When sealing the bottles use of a cotton plug

4) Harm from mites

There is a certain possibility of spreading mites from the external environment. To avoid that apply a mild pesticide once in 6 months.

References

1. Bhojwani, Saran, S., Prem Kumar, D., (2013). Plant Tissue Culture: An Introductory Text. Springer Publications.

2. Micropropagation techniques. [online] Available at: <http /biotechnology/clonal-propagation/micro-propagation-technique-factors-applications-and-disadvantages/10732 > [Accessed 10 July 2017].

3. Ogero, K. O., Gitonga, N. M., Mwangi, M., Ombori, O. and Ngugi, M. (2012) 'Cost-effective nutrient sources for tissue culture of cassava (Manihot esculenta Crantz)', 11(66), pp. 12964–12973. doi: 10.5897/AJB12.579.

4. Production of Synthetic Seed by Encapsulating Asexual Embryo in Eggplant (Solanum melongena L.) Available <http://scialert.net/fulltext/?doi=ijar.2007.832.837> [Accessed 15 June 2017].

5. Saad, A. I. M. and Elshahed, A. M. (2012) 'Plant Tissue Culture Media': Recent Advances in Plant *in vitro* Culture. Edited by Leva,A. and Rinaldi, L.M.R. Intech.

6. Smith, R. (2012). Plant Tissue Culture Techniques and Experiments. 3rd ed. Academic Press.

yes
I want morebooks!

Buy your books fast and straightforward online - at one of world's fastest growing online book stores! Environmentally sound due to Print-on-Demand technologies.

Buy your books online at
www.morebooks.shop

Kaufen Sie Ihre Bücher schnell und unkompliziert online – auf einer der am schnellsten wachsenden Buchhandelsplattformen weltweit! Dank Print-On-Demand umwelt- und ressourcenschonend produzi ert.

Bücher schneller online kaufen
www.morebooks.shop

Printed by Books on Demand GmbH, Norderstedt / Germany